Multiple-choice Questions in Basic Sciences for the MRCPsych Part II

Other Examination Preparation Books Published by Petroc Press:

Multiple-choice Questions in Basic Sciences for the MRCPsych Part II

Sudip Sikdar, MBBS, MD, MRCPsych

Specialist Registrar in Psychiatry
Fazakerley Hospital, Longmoor Lane, Liverpool L9 7AL, UK

CRC Press
Taylor & Francis Group
Boca Raton London New York

CRC Press is an imprint of the
Taylor & Francis Group, an **informa** business

First published 1999 by CRC Press

Published 2019 by CRC Press
Taylor & Francis Group
6000 Broken Sound Parkway NW, Suite 300
Boca Raton, FL 33487-2742

© 1999 by Taylor & Francis Group, LLC
CRC Press is an imprint of the Taylor & Francis Group, an informa business

No claim to original U.S. Government works

ISBN-13: 978-1-138-44801-8 (hbk)
ISBN-13: 978-1-900603-56-0 (pbk)

**Visit the Taylor & Francis Web site at
http://www.taylorandfrancis.com**

**and the CRC Press Web site at
http://www.crcpress.com**

Contents

Preface

The idea of this book was forming when I was writing the Part II examination. I could not find a single comprehensive book that covered solely the basic sciences part of the examination, and hence I struggled greatly preparing it. Though there are standard recommended textbooks for the clinical paper, there is no such common ground for the basic sciences paper. To make matters worse, the examination does not, as yet, fully reflect the curriculum, some topics being examined in detail while others are covered only superficially if at all. Rather than mixing up different questions from different chapters of the syllabus, I have deliberately kept them separate, hoping that it will help in the revision of particular chapters. The number of questions in each chapter is intended to reflect the importance of the topic in the examination. I have also deliberately kept the explanatory answers brief, with the intention of encouraging the student to read in greater depth around the topics in question. Because of the general vagueness of the basic sciences paper, it is extremely difficult to reflect exactly the real examination questions, and I must admit that this book does not attempt to do the impossible. However, I hope that it will help in preparation for this difficult part of the examination.

Liverpool, 1998   S.S.

1. Human Development

Q1.1 A child of 3 years:

A. Can ride a tricycle
B. Can give his/her full name and sex
C. Can copy a cross
D. Can put on shoes
E. Can name at least one colour

Q1.2 The following statements are true about attachment behaviour:

A. Attachment of mothers to infants occurs crucially during the early hours following birth
B. Studies have failed to find an increase in the strength of mothering behaviour as a result of extended contact with the infant
C. Stranger anxiety is a necessary component of attachment behaviour
D. Onset of attachment behaviour coincides with development of object permanence
E. Avoidant attachment in childhood is often associated with antisocial behaviour in adult life

Q1.3 The following statements are true about neonatal behaviour:

A. Rudimentary stepping behaviour in neonates disappears by 2 months
B. Truly intentional social behaviour appears around 6 months
C. Truly reciprocal social behaviour appears around 6 months
D. *Motherese* refers to modification of a mother's facial expression when interacting with her child
E. Adult level of vision is reached by 8 months

Q1.4 The following statements are true of Thomas and Chess's NYLS (New York Longitudinal Study) of temperament:

A. It was based on the assumption that temperament is largely genetically determined
B. It used the Child Behaviour Checklist to formulate the dimensions of temperament
C. It recruited the sample from a homogeneous middle-class group of parents
D. Activity levels accounted for the greatest variance between children
E. Slow to warm up babies were predisposed to developing later behaviour disorder

1

Q1.5   'Resilience in the face of risk' was attributed by Thomas and Chess to:

A. Development of schizoid personality in adult life
B. Development of commitment to a career
C. Distancing from parents as a young person
D. Brain damage
E. A good relationship with someone outside the family

Q1.6   In Piaget's stage of precausal logic, a child can:

A. Reason based on observations
B. Accept false explanations
C. Imitate
D. Detach words from the objects they symbolise
E. Reason about quantity

Q1.7   The following are true about language development in children:

A. Rate of language development is linked with intelligence
B. Skinner proposed that acquisition of language is acquisition of a rule system and a code
C. Bruner saw the basis of grammar in preverbal social exchange between mother and child
D. Chomsky demonstrated that children can copy adult grammar
E. Children become proficient in language because of what they hear others say

Q1.8   The following statements are true about adolescence:

A. Absence of adolescent turmoil indicates maladaptation
B. The key task of adolescence is development of personal identity
C. Identity confusion is a source of serious turmoil in normal teenagers
D. Most adolescents identify with their parents' basic moral principles
E. Social phobia has a special relationship with adolescent development

Q1.9   The following statements on genetic factors on development are correct:

A. There is a heritable component in crimes against property
B. There is a strong genetic loading in juvenile delinquency
C. Twin and adoption studies support a strong genetic contribution in petty criminality
D. A neurobiological correlate for harm avoidance has been found
E. The behavioural characteristics of distractibility has a strong genetic component

Q1.10 The following statements are true regarding development of empathy in children:

A. Newborns do not have the ability to respond to the feelings of others
B. True sympathy develops in the second year of life
C. Reflexive emotional resonance starts after the child develops the ability to distinguish between his or her own feelings and those of others
D. The empathic responses become attuned to needs of others from the third year onwards
E. According to Hoffmann, development of empathic responsiveness is dependent on the prospects of reward and punishment by the parents

Q1.11 According to Bentovim, successful families should include:

A. A model for socialisation
B. Models in the parents for sexual identity
C. Mutual dependency and investment
D. Triangular relationship structures
E. Boundaries demarcating parents and children

Q1.12 According to West and Farrington, the following factors predict criminality:

A. Low intelligence
B. Criminality in the father
C. Criminality in the mother
D. Families with more boys than girls
E. Poor parental conduct

Q1.13 Names associated with attachment theory include:

A. Harlow
B. Rutter
C. Lorenz
D. Kliene
E. Winnicot

Q1.14 The following statements are true about the development of antisocial behaviour:

A. Twin and adoption studies have reported a strong genetic influence in juvenile delinquency
B. There is research evidence to show that antisocial behaviour is more closely related to social class than with environmental influence
C. Recent studies have shown a direct relationship between delinquency and family size
D. Family discord is more important than family breakdown as a predictor of future delinquency
E. Multiple regression techniques are better predictors of delinquency than linear methods

Q1.15 The following are true about ethological studies:

A. Tinbergen coined the term *ethology*
B. Imprinting can occur only during a brief period after birth
C. Imprinting affects later sexual behaviour
D. Japanese orphans brought up in China developed Chinese mannerisms
E. Chinese orphans brought up in Japan retained Chinese mannerisms

1. Answers

A1.1
A. T
B. T
C. F It takes 4 years for a child to achieve this milestone
D. T
E. F It takes 4 years for a child to achieve this milestone

A1.2
A. F There is no such time restriction, although this phenomenon has been termed *bonding* by some
B. T
C. F It is not, but many assume that it is
D. T
E. T

A1.3
A. T
B. F Truly intentional social behaviour appears around 3 months
C. T
D. F *Motherese* refers to modification of the mother's speech when interacting with her child
E. F Adult level of vision is reached by 6 months

A1.4
A. F Temperament is an interactional concept between gene and environment
B. F It used the NYLS Parent Questionnaire
C. T
D. F Approach/withdrawal accounted for the greatest variance
E. F Slow to warm up babies are predisposed to developing phobic disorder

A1.5
A. F It was attributed to development of a gift in middle childhood
B. T
C. T The distancing was sometimes geographical
D. F Poor outcome despite low risk factors was found in those with brain damage, depressive illness or schizoid personality
E. T

A1.6
A. **F** Reasoning is based on the child's internal model of the world
B. **T**
C. **T**
D. **F** A child cannot detach words from the objects they symbolise
E. **T**

A1.7
A. **T**
B. **T**
C. **T**
D. **F** They cannot, but invent their own grammatical rules
E. **F** They become proficient as a result of encouragement to speak

A1.8
A. **F** It indicates health
B. **F** Most adolescents are content to adopt the identity ascribed to them by their family of origin
C. **F** It is a common issue among disturbed adolescents
D. **T**
E. **T**

A1.9
A. **T**
B. **F** The genetic component is of much lesser magnitude
C. **F** The genetic contribution in petty criminality is weak
D. **T**
E. **F** Distractibility has a weak genetic component

A1.10
A. **F** Newborns do have the ability to respond to the feelings of others
B. **T**
C. **F** Newborns have this capacity
D. **T**
E. **F** Is independent of prospects of reward and punishment

A1.11
A. **T**
B. **T**
C. **F**
D. **F**
E. **T**

A1.12
A. T
B. T
C. T Criminality in the mother is more predictive than criminality in the father
D. F There is no such relationship
E. T

A1.13
A. T
B. F
C. T
D. F
E. F

A1.14
A. F Only moderate genetic influence has been found in antisocial behaviour
B. F Antisocial behaviour is more closely related to environmental influence than social class
C. F There is only an assumed link
D. T
E. F Multiple regression gives poorer prediction

A1.15
A. F Lorenz did
B. T
C. T
D. F They retained Japanese mannerisms because of the enduring effects of early parent–child interaction with their Japanese parents
E. F No such studies have been done

2. Genetics

Q2.1   The following statements about the human genome are correct:

A. In a normal human genome, about 50% of total genomic DNA is non-coding
B. It is estimated that 30–50% of the human genome is expressed mainly in the brain
C. Mitochondrial chromosomes are identical to nuclear chromosomes
D. During mutation, unstable complementary DNA is produced
E. DNA separates its double helix in a reaction catalysed by reverse transcriptase

Q2.2   The following statements about human DNA are correct:

A. Mammalian DNA is supercoiled around histone proteins
B. tRNA has an anticodon at one end to attach to an mRNA
C. Histone proteins are not transcribed
D. Satellite DNAs are inherited in a non-Mendelian fashion
E. cDNA is produced by reverse transcriptase

Q2.3   The following diseases and their abnormal gene locations are correctly paired:

A. Huntington's disease : 4q 16.3
B. Fragile X syndrome : Xp 27.3
C. Lesch–Nyhan syndrome : Xq 26
D. Wilson's disease : 13q 14
E. Friedreich's ataxia : 9q 12

Q2.4   In disorders with autosomal dominant transmission:

A. The phenotypic trait is present in all individuals carrying the dominant allele
B. When one parent is homozygous, half the offspring will manifest the abnormal trait
C. When a normal individual mates with a heterozygote individual, three quarters of the offspring will manifest the abnormal trait
D. When two heterozygotes mate, all the offspring will manifest the abnormal trait
E. Male and female offspring have equal risk of being affected

Q2.5   In disorders with autosomal recessive transmission:

A. Only double heterozygotes manifest the abnormal phenotypic trait
B. Single heterozygotes are carriers of the abnormal trait
C. When two heterozygotes mate, half the offspring will be affected
D. When an affected individual mates with a normal one, a quarter of the offspring will be affected
E. Male and female offspring have an equal risk of being affected

Q2.6   In disorders with X-linked recessive transmission:

A. Only male offspring will manifest the abnormal phenotypic trait
B. When an affected male mates with a normal female, half the daughters will be carriers
C. When an affected female mates with a normal male, all the daughters will be carriers
D. When a carrier female mates with a normal male, half the daughters will be carriers
E. Half the female heterozygotes are carriers

Q2.7   The following statements are true about twin studies in psychiatry:

A. Monozygotic twins have a more dissimilar environmental influence than dizygotic twins
B. The concordance rate in dizygotic twins is higher than in monozygotic twins when they are born out of wedlock
C. The concordance rate in monozygotic twins is higher than in dizygotic twins in parental assortative mating
D. Pairwise concordance gives lower rates than probandwise concordance
E. Taking a hospitalised sample is the best way to study twin concordance

Q2.8   The following statements are true about adoption studies in psychiatry:

A. In adoptee family studies, illness rates are compared in the biological and adoptive parents of normal adoptees
B. In an adoptee study, illness rates are compared in the biological and adoptive parents of ill adoptees
C. In cross-fostering studies, illness rates are compared in fostered and normal children
D. Most cases for adoption studies are taken from the adoption register
E. The process of adoption is non-random

Q2.9 The following statements are correct:

A. A gene is said to have variable expressivity when it causes varying manifestation of an exophenotype depending on variations in environmental factors
B. Single gene defects are inherited in a non-Mendelian fashion
C. Heritability of a trait is defined by the frequency of expression of a dominant gene
D. Penetrance of a trait is defined as the proportion of total phenotypic variance contributed by the genetic component
E. Characters that are polygenically inherited do not have a continuous population distribution

Q2.10 The following statements are correct:

A. A codon is a sequence of three nucleotide bases that can code for an amino acid
B. In the somatic cell nucleus, each gene exists in an allelic pair
C. During the process of hybridisation, complementary DNA strands are separated
D. A recombinant DNA is derived from more than one organism
E. Cloning involves insertion of a DNA fragment into a cosmid capable of autonomous replication in a host cell

Q2.11 A gene probe:

A. Is a fragment of cDNA
B. Can be constructed by the enzyme restriction endonuclease
C. Can be constructed from mRNA
D. Has a base sequence complementary to that of a given part of a genome
E. Is specifically related to the disease gene for most diseases

Q2.12 Restriction fragment length polymorphism (RFLPs):

A. Are inherited in a simple Mendelian pattern
B. Can be used as DNA markers
C. Detect very short sequences of repeated dinucleotides
D. Are DNA fragments of different lengths in an individual
E. Can be used as gene probes

Q2.13 The following statements about recombination in genetics are correct:

A. The closer two loci are during meiosis, the higher are their chances of recombination
B. Recombinant function measures the frequency of separation of alleles during meiosis
C. The value of recombinant function varies between 0 and 1
D. Linkage measures the frequency of inheritance of two genes in close proximity
E. Lod scores are most effective for conditions with a non-Mendelian mode of inheritance

Q2.14 The following statements about association studies in molecular genetics are true:

A. They are in most respects more complicated than linkage studies
B. They compare the frequency of a particular marker gene in patients and in healthy controls
C. In pleiotropy, different genes apparently affect the same phenotypes
D. In linkage disequilibrium, the recombination fraction is very low
E. Association studies enable detection of genes accounting for a very small proportion of variance in liability

Q2.15 In the liability threshold model of transmission of psychiatric disorder:

A. It is supposed that there is an underlying graded genetic liability to develop the disorder
B. In the general population, plotting the frequency of the disorder against its liability yields a normal distribution
C. In patients suffering from a particular disorder, the curve is shifted to the left
D. The aim of segregation analysis is to find out how much of the environmental influence is due to shared and how much to non-shared influence
E. The sib-pair method aims to elucidate the mode of transmission of a trait rather than simply to estimate its variance component

2. Answers

A2.1
A. F 90% of genomic DNA is non-coding
B. T
C. F There are two copies of nuclear chromosomes in every cell, but there are numerous mitochondrial chromosomes
D. F mRNA is produced
E. F DNA polymerase catalyses DNA separation

A2.2
A. T
B. T
C. T
D. F Satellite DNAs are inherited in a Mendelian fashion
E. T

A2.3
A. F 4p 16.3
B. T
C. T
D. T
E. T

A2.4
A. T
B. F All offspring would be affected
C. F Half will manifest it
D. F Three quarters will manifest it
E. T

A2.5
A. F Only homozygotes manifest it
B. T
C. F A quarter will be affected
D. F None will be affected
E. T

A2.6
A. F None of the sons will be affected when an affected male mates with a normal female
B. F All the daughters will be carriers
C. T
D. T
E. F All female heterozygotes are carriers

A2.7
A. T Due to competing effect with the co-twin
B. F
C. F The reverse is true
D. T
E. F Twin concordance is best studied from a twin register, as hospitalised samples contain extremely ill twins, which may give falsely higher concordance rates

A2.8
A. F
B. F
C. F
D. F Adoption registers are not commonly used
E. T An effort is made to find parents similar in character to adoptees

A2.9
A. F Such variations are independent of environmental factors
B. F Single gene defects are inherited in a Mendelian pattern
C. F This is the penetrance
D. F This is the heritability
E. F They do have continuous distribution

A2.10
A. T
B. T
C. F They are brought together; strands are separated in denaturation
D. T
E. T

A2.11
A. T
B. T
C. T By reverse transcriptase
D. T
E. F It is only linked to the disease gene, not specifically related

A2.12
A. T
B. T
C. T
D. F They are DNA fragments of similar lengths
E. F

A2.13
A. F
B. T
C. F It varies between 0 and 0.5
D. T
E. F Lod scores are most effective for conditions with Mendelian inheritance

A2.14
A. F They are easier than linkage studies
B. F They compare marker phenotypes
C. F The same gene affects two or more different phenotypes
D. T Because the disease and marker genes are too close to each other
E. T

A2.15
A. T
B. T
C. F It is shifted to the right
D. F The aim is to elucidate the mode of transmission of a trait rather than simply to estimate its variance component
E. F It is a more robust method of linkage analysis

3. Neurosciences

Q3.1   In cranial nerve palsies:

A. One of the first signs of a compressive pathology in the pituitary region is a bitemporal field defect
B. Progressive failure of gaze in all directions may be associated with Parkinson's syndrome
C. Lesions in the cerebellopontine angle cause loss of corneal sensation
D. Gaze is directed towards the paralysed side in lesions affecting the frontal lobe
E. The disc margins appear raised in hypermetropic eyes on fundoscopy

Q3.2   By making a patient stand with eyes shut and hands out-stretched, one can check:

A. Pyramidal function
B. Extrapyramidal function
C. Parietal function
D. Frontal function
E. Cerebellar function

Q3.3   In the above posture:

A. The palms must face upwards
B. Minimal pyramidal loss causes only spreading of the fingers
C. Drifting of the whole hand with an abnormal posture is a sign of extra-pyramidal disease
D. Cerebellar disease causes the arm to drift upwards and lose position
E. In parietal lesions, the posture of the hand may be abnormal and continuously changing

Q3.4   The following are limbic pathways:

A. Stria terminalis
B. Stria medullaris
C. Median forebrain bundle
D. Medial longitudinal stria
E. Dorsal longitudinal stria

Q3.5   The following statements about brain structures are true:

A. The striatum is made up of the caudate nucleus and the globus pallidus
B. The pars reticulata uses dopamine as its main neurotransmitter
C. Basal ganglia output nuclei exert cholinergic inhibition on the thalamus
D. During development, the first brain structure to myelinate is the pre-frontal cortex
E. The whole of the nervous system develops from the ectoderm of the embryo

Q3.6   The following are tests of frontal lobe function:

A. Figure fluency
B. Cognitive estimation
C. Multiple loops test
D. Three step hand sequence
E. Proverb interpretation

Q3.7   The following relationships between brain structures and functions are correct:

A. Destruction of the ventromedial hypothalamus : hyperphagia
B. Limbic system activation : kindling
C. Septohippocampal system : anxiety modulation
D. Amygdala : memory for emotional events
E. Anterior hypothalamic lesion : activation of sexual activity

Q3.8   The following are potential hypothalamic peptide neurotransmitters:

A. Bombesin
B. Bradykinin
C. Secretin
D. Angiotensin
E. α-Endorphin

Q3.9   The following neurotransmitters and peptides coexist:

A. Dopamine : encephalin
B. Noradrenaline : encephalin
C. Acetylcholine : CCK (cholecystokinin)
D. Serotonin : TRH (thyrotropin-releasing hormone)
E. Adrenaline : encephalin

Q3.10 In Huntington's chorea:

A. The brain is normal in size
B. The frontal lobes degenerate
C. GAD levels are low
D. The average age of onset is in the mid-twenties
E. Insight is lost early

Q3.11 The following statements about cerebral tumours are true:

A. Adults suffer mainly from supratentorial tumours
B. Ependymomas spread via the cerebrospinal fluid
C. Meningiomas grow rapidly
D. Medulloblastomas are the commonest primary tumours in childhood
E. Acoustic neuromas affect cranial nerves V, VI, VII and VIII

Q3.12 In Parkinson's disease:

A. Depigmentation is seen mainly in the zona reticulosa
B. Lewy bodies are seen in the dead neurons
C. Cortical atrophy is rare
D. The cranial nerves are spared
E. The reticular formation is affected

Q3.13 The following statements about cerebral inclusion bodies are true:

A. Pick bodies are argyrophilic extracytoplasmic neuronal inclusions
B. Lewy bodies are hyaline intracytoplasmic neuronal inclusions
C. Hirano bodies are eosinophilic intracytoplasmic neuronal inclusions
D. Opalski cells are found in Hunter's syndrome
E. Zebra bodies are found in Hurler's disease

Q3.14 In multi-infarct dementia:

A. The degree of cognitive impairment correlates with the extent of infarction
B. Features of Klüver–Bucy syndrome may be seen
C. In contrast to Alzheimer's disease, the sex ratio is equal
D. A minimum of 100 ml of infarction is required before cognitive impairment is detectable
E. A volume of 100 ml is particularly likely to be associated with dementia

Q3.15 Neurofibrillary tangles are seen in:

A. Amyotrophic lateral sclerosis
B. Down's syndrome
C. Pick's disease
D. Punch-drunk syndrome
E. Dementia of frontal lobe type

Q3.16 In Punch-drunk syndrome:

A. The lateral ventricles are commonly enlarged
B. Cerebral atrophy is unusual
C. Neuritic plaques are sometimes visible
D. Confabulation is a feature found commonly
E. The corpus callosum is perforated

Q3.17 Demyelination occurs in the following conditions:

A. Schilder's disease
B. Niemann–Pick disease
C. Adrenoleucodystrophy
D. Gaucher's disease
E. Tay–Sach's disease

Q3.18 In Wernicke–Korsakoff syndrome:

A. The pathological changes are asymmetrical
B. Demyelination is seen
C. The cerebellum is spared
D. Cerebral atrophy is associated with Wernicke's disease
E. Ventricular dilatation may be seen in Wernicke's disease

Q3.19 Parkinsonism may be caused by:

A. Mercury poisoning
B. Carbon dioxide poisoning
C. Lead poisoning
D. Cerebral palsy
E. Wilson's disease

Q3.20 The new variant of CJD:

A. Presents predominantly with symptoms of dementia
B. Has a course that is slower than that of typical CJD
C. Has scanty PrP amyloid plaques
D. Shows type 4 PrP bands on electrophoresis
E. Has only valine at codon 129

Q3.21   The following statements about evoked potentials are true:

A. The initial waves in an evoked potential tracing are dependent on the psychological state of the individual

B. The P300 wave is said to relate to a process of cognitive appraisal of the stimulus

C. Contingent negative variation (CNV) is another name for 'readiness potential'

D. In schizophrenia, the amplitude of the P300 wave is usually increased

E. The latency of the P300 wave is significantly delayed in dementias

Q3.22   In normal sleep:

A. After intense physical exercise, the proportion of slow wave sleep increases

B. After a lengthy period of wakefulness, the proportion of REM sleep increases

C. Very little stage 2 sleep is regained following a period of sleep deprivation

D. Stage 3 sleep predominates in optional sleep

E. Few sleep spindles are seen in stage 4 sleep

Q3.23   The following statements are true of a normal EEG:

A. Beta and theta rhythms dominate at birth

B. Alpha rhythm is established around 6 years

C. Posterior temporal theta activity is not infrequently seen in younger people

D. Paroxysmal focal spikes are common in children

E. Only a limited number of channels can be recorded in ambulatory EEG

Q3.24   The following statements are true of normal sleep:

A. Delta rhythm is predominant in stage 3

B. It is easier to wake people from REM than from NREM sleep

C. More than 50% of dreams in REM sleep can be remembered

D. There is no evidence to suggest that eye movements are related to dream content during REM sleep

E. The biological effect of total sleep deprivation is severe in the stress-related endocrine system

Q3.25 The following statements are true of EEG:

A. Electrical activity can be measured as early as 12 weeks in the human foetus
B. Immaturity is defined as the presence of an excess of slow waves for the age
C. During ageing, alpha rhythm is better preserved in men
D. Lithium can produce bursts of theta activity
E. Opiates produce little EEG change when taken by addicts

Q3.26 The following statements about neuroreceptors are true:

A. D5 receptors inhibit adenylate cyclase
B. D5 receptors are predominantly seen in the prefrontal cortex
C. D1 receptor activation can enhance intracellular D2 receptor activity
D. $5HT_3$ receptors are linked to adenylate cyclase
E. Stimulation of 5HT receptors decreases acetylcholine release

Q3.27 The following statements about neurophysiological studies of the brain are correct:

A. In MRI studies, T1 relaxation time is always greater than T2 relaxation time
B. T1 relaxation time relates to interaction of proton molecules with each other
C. Image resolution in magnetic resonance spectroscopy is less than in MRI
D. A patient cannot have more than one functional MRI scan at a time
E. SPECT studies require an on-site cyclotron

Q3.28 The following statements about synaptic transmission are correct:

A. The primary structure is similar for all ion channels
B. Amplification is an important feature of the inositol phosphate second messenger system
C. All postsynaptic receptors have five subunits in their general structure
D. The amount of neurotransmitter released during an action potential is related to the calcium levels at the presynaptic terminal
E. Inhibitory postsynaptic potential results from an influx of potassium and chloride

Q3.29 The following statements about prolactin are correct:

A. Prolactin levels tend to be elevated in patients receiving intramuscular therapy
B. Prolactin levels tend to be elevated in patients receiving intravenous therapy
C. Prolactin levels correlate poorly with serum levels of neuroleptics
D. Prolactin levels are predictors of dopamine blockade at the nigrostriatal axis
E. β-Endorphins stimulate prolactin release

Q3.30 The neural crest gives rise to:

A. The dorsal root ganglia
B. The cerebral grey matter
C. The spinal nerves
D. The cranial nerves
E. The adrenal cortex

Q3.31 The following arteries are branches of the internal carotid artery:

A. Posterior choroidal
B. Posterior cerebral
C. Ophthalmic
D. Labyrinthine
E. Anterior choroidal

Q3.32 The following substances pass through the blood–brain barrier by simple diffusion:

A. Carbon dioxide
B. Water
C. Alcohol
D. Glucose
E. Amino acids

Q3.33 The thalamus:

A. Is part of the telencephalon
B. Receives input from all the senses
C. Lateral geniculate nucleus receives information from the ears
D. Anterior nucleus receives taste sensation
E. Dorsomedial nucleus receives general sensation

Q3.34 Infarction in the territory of the anterior cerebral artery leads to:

A. Aphasia
B. Ipsilateral Horner's syndrome
C. Contralateral sensory loss
D. Contralateral hemiplegia
E. Clouding of consciousness

Q3.35 The following types of nystagmus are correctly paired with their causative lesions:

A. Horizontal : middle ear disease
B. Horizontal : cerebellar disease
C. Vertical : multiple sclerosis
D. Vertical : brain stem disease
E. Ataxic : phenytoin overdose

Q3.36 The principal outputs of the basal ganglia go to:

A. The red nucleus
B. The tectum
C. The cerebral cortex
D. The subthalamic nucleus
E. The substantia nigra

Q3.37 Diplopia may occur in:

A. Neuropathy of the optic nerve
B. Parkinson's disease
C. Huntington's chorea
D. Wilson's disease
E. Neuropathy of the oculomotor nerve

Q3.38 Serotonergic cell bodies are found in:

A. The substantia nigra
B. The dorsal raphe nucleus
C. The pontomedullary region of the brain stem
D. The corpus callosum
E. The median raphe nucleus

Q3.39 The following statements are true about glycine as a neuro-transmitter:

A. It is predominantly found in the brain
B. Its principal effect is on the presynaptic membrane
C. It is a weak excitatory neurotransmitter
D. It hypopolarises motor neurons
E. Strychnine competes for its receptors

Q3.40 In the brain:

A. The lipid content is minimal
B. Gangliosides are the predominant type of lipid
C. Less than 1% of cholesterol is in the free form
D. The brain lipids are relatively unaffected by dietary lipids
E. Triglycerides and free fatty acids are abundant

Q3.41 The following statements regarding substance P are correct:

A. It consists of 11 amino acids
B. A high concentration is found in the cortex
C. There is a large nigrostriatal pathway involving it in the brain
D. It is most likely to be an inhibitory neurotransmitter
E. It can cause both hyperalgesia and analgesia

Q3.42 The following receptors are cation channel linked:

A. Nicotinic
B. Glycine
C. $5HT_3$
D. $5HT_2$
E. Delta

Q3.43 The following receptors are G protein linked:

A. GABA-B
B. Dopamine D1 and D2
C. $5HT_{1C}$
D. $5HT_2$
E. Muscarinic

Q3.44 The following components of language are correctly related to their anatomical structures:

A. Phonology : left superior temporal lobe

B. Semantics : left temporal lobe

C. Syntax : left anterior hemisphere

D. Prosody (fine tuning) – right hemisphere

E. Prosody (emotional expression) – left anterior hemisphere

Q3.45 The following neuropathological features of Parkinson's disease are correct:

A. Pallor of the locus ceruleus

B. Pallor of Meynert's nucleus

C. Pallor of the Edinger–Westphal nucleus

D. Pallor of the sympathetic nuclei

E. Pallor of the vagus nucleus

Q3.46 The following inhibit growth hormone secretion:

A. Increased free fatty acids

B. Obesity

C. Hepatic cirrhosis

D. Insomnia

E. Renal failure

Q3.47 The following are found in sociopaths:

A. Increased cortical arousal

B. Decreased skin conductance

C. Exaggerated response to stress

D. Rapid development of conditioning to fear-provoking stimuli

E. Diffuse slow wave on EEG

Q3.48 The following stimulate prolactin secretion:

A. Cushing's disease

B. Renal failure

C. Empty sella syndrome

D. Secondary hypothyroidism

E. Hepatic cirrhosis

Q3.49 The following signs are correctly paired with their site of origin:

A. Anosognosia : dominant parietal
B. Prosopagnosia : non-dominant parietal
C. Alexia with agraphia : dominant temporal lobe
D. Complex visual hallucination : non-dominant occipital lobe
E. Astereognosis in the right hand : corpus callosum

Q3.50 Hyperprolactinaemia is associated with:

A. Paedophilia
B. Ejaculatory failure
C. Erectile failure
D. Klinefelter's syndrome
E. Libidinal failure

Q3.51 The following statements are true about the GABA shunt:

A. It is unique to the brain
B. It accounts for a quarter of the total glucose turnover in the brain
C. It is a bypass around the tricarboxylic acid cycle from ketoglutarate to succinate
D. It arises because of the presence of the enzyme glutamate transaminase
E. It requires vitamin B6 as a cofactor

Q3.52 The following statements are true about biochemical markers:

A. DNA can be used as an index of cell numbers
B. Cholesterol can be used as an indicator of myelin
C. DNA can be used as an indicator of cell size
D. Acetylcholinesterase can be used as an indicator of synapses
E. Gangliosides can be used as indicators of glial cells

Q3.53 SPECT:

A. Employs radiochemicals that emit positrons
B. Has better image resolution in than positron emission tomography (PET)
C. Can measure metabolism in different areas of brain in one individual
D. Does not have any radiation risk
E. May be used to compare cerebral metabolic measurements within patient groups

3. Answers

A3.1
A. **F** Bitemporal loss of red colour appreciation is one of the first signs
B. **T** When associated with progressive supranuclear palsy
C. **T**
D. **F** It is directed towards the side of the lesion
E. **T**

A3.2
A. **T**
B. **F**
C. **T**
D. **F**
E. **T**

A3.3
A. **F** They may face downwards
B. **T**
C. **F** This indicates severe pyramidal disease
D. **F** Parietal disease causes this
E. **F** Cerebellar disease causes this

A3.4
A. **T**
B. **T**
C. **F** The medial forebrain bundle is a limbic pathway
D. **T**
E. **F** The dorsal longitudinal *fasciculus* is a limbic pathway

A3.5
A. **F** It consists of the caudate and the putamen
B. **F** Its main neurotransmitter is GABA; the pars compacta uses dopamine
C. **F** They exert GABAergic inhibition
D. **F** This is the last structure to myelinate
E. **T**

A3.6

A. T
B. T
C. T
D. T
E. T

A3.7

A. T
B. T
C. T
D. T
E. F Anterior hypothalamic lesions are associated with prevention of sexual activity

A3.8

A. T
B. T
C. T
D. T
E. F β-endorphins are potential neurotransmitters

A3.9

A. T
B. T
C. F Acetylcholine and VIP (vasoactive intestinal polypeptide) coexist
D. T
E. T

A3.10

A. F The brain is small in size
B. F
C. T
D. F Average onset is in the thirties
E. F Insight is retained

A3.11

A. T
B. T
C. F Meningiomas are usually slow growing
D. T
E. T

A3.12
A. T
B. F Lewy bodies are seen in surviving neurons
C. F Diffuse cortical atrophy occurs
D. F The dorsal vagal nerve nucleus is affected
E. T

A3.13
A. F Pick bodies are intracytoplasmic
B. T
C. T
D. F Opalski cells are seen in Wilson's disease
E. T

A3.14
A. T
B. F These are seen in Alzheimer's disease
C. F Males are more commonly affected
D. F A minimum of 50 ml of infarcted brain tissue is required before cognitive impairment is detected
E. T

A3.15
A. T
B. T
C. F
D. T
E. F

A3.16
A. T
B. F Gross atrophy occurs
C. F Plaques are not seen
D. F Confabulation is not found
E. F The septum pellucidum is perforated

A3.17
A. T
B. T
C. T
D. T
E. T

A3.18
A. F The changes are bilaterally symmetrical
B. T
C. F The superior vermis of the cerebellum is affected
D. F Atrophy is seen only in Korsakoff's syndrome
E. T

A3.19
A. T
B. F Carbon monoxide poisoning can cause parkinsonism
C. T
D. T
E. T

A3.20
A. F New variants of CJD present predominantly with psychiatric symptoms
B. T
C. F Numerous PrP amyloid plaques are seen
D. T
E. F All the amino acids at codon 129 are methionine

A3.21
A. F They are independent of psychological state
B. T
C. F Readiness potential is a positive motor potential arising 1 second before voluntary movement
D. F The amplitude of the P300 wave is decreased in schizophrenia
E. T

A3.22
A. T
B. F Slow wave sleep increases after lengthy wakefulness
C. T
D. F Stage 2 sleep predominates in optional sleep
E. F Sleep spindles are no longer seen in stage 4 sleep

A3.23
A. F Delta rhythm dominates at birth
B. F Alpha rhythm is established around 13 years
C. T
D. F They are seen in only 3% of children
E. T

A3.24
A. **F** Delta sleep is predominantly seen in stage 4
B. **F** The opposite is true
C. **T**
D. **F** Some evidence is emerging
E. **F** Surprisingly, it is extremely limited

A3.25
A. **F** Electrical activity arises at 20 weeks in a human foetus
B. **T**
C. **F** It is better preserved in women
D. **F** Lithium produces delta rhythm, which may be focal
E. **T**

A3.26
A. **F** D5 receptors stimulate adenylate cyclase
B. **F** D4 receptors are predominantly seen in the prefrontal cortex
C. **T**
D. **F** $5HT_3$ receptors are linked to ion channels
E. **T**

A3.27
A. **T**
B. **F** T1 relaxation time relates to interaction of protons with surrounding nuclei
C. **T**
D. **F** Multiple scans are possible as it is non-invasive and does not require radioactive substances
E. **F** They do not, as the tracers used have a long half-life

A3.28
A. **T**
B. **F** Amplification is an important feature of the adenylate cyclase system because each activated receptor protein stimulates many G protein molecules, which in turn activate many molecules of adenylate cyclase, each generating many cAMP molecules
C. **T**
D. **T**
E. **T**

A3.29

A. T

B. F

C. T

D. F Prolactin levels predict dopamine blockade at the hypothalamo-pituitary axis

E. T

A3.30

A. T

B. F The neural tube gives rise to the cerebral grey matter

C. T

D. T

E. F The neural crest gives rise to the adrenal medulla

A3.31

A. F This is a branch of the posterior cerebral artery

B. F This is a branch of the basilar artery

C. T

D. F This is a branch of the basilar artery

E. T

A3.32

A. T

B. T

C. T

D. F

E. F Glucose and amino acid need facilitated transport

A3.33

A. F The thalamus is part of the diencephalon

B. F It receives no input from the sense of smell

C. F It receives information from the eyes

D. F The ventral posterior nucleus receives taste

E. F The ventral posterior nucleus receives general sensation

A3.34

A. T

B. T

C. T

D. T

E. T

A3.35
A. T
B. T
C. F
D. T
E. F   Ataxic nystagmus is highly suggestive of multiple sclerosis

A3.36
A. T
B. T
C. F
D. T
E. T

A3.37
A. F   It occurs in neuropathy of the oculomotor nerve
B. F
C. F
D. F
E. T

A3.38
A. T
B. T
C. T
D. F
E. T

A3.39
A. F   Glycine is predominantly found in the spinal cord
B. F   Its principal effect is on the postsynaptic membrane
C. F   It is a potent inhibitory neurotransmitter
D. F   It hyperpolarises postsynaptic membrane
E. T

A3.40
A. F   The brain is rich in lipids
B. F   Cholesterol is the predominant lipid in the brain
C. F   Most of the cholesterol exists in the free form
D. T
E. F   Very little is found in the brain

A3.41

A. T

B. F High levels are found in the basal ganglia and hypothalamus

C. T

D. F It is an excitatory neurotransmitter in pain signalling and for certain dopaminergic pathways

E. T

A3.42

A. T These receptors are sodium channel linked

B. F These receptors are chloride channel linked

C. T These receptors are sodium channel linked

D. F These receptors are phosphoinositol linked

E. T These receptors are potassium channel linked

A3.43

A. T

B. T

C. F These receptors are phosphoinositol linked

D. F These receptors are phosphoinositol linked

E. T

A3.44

A. T

B. T

C. T

D. F Left anterior hemisphere

E. F Right hemisphere

A3.45

A. T

B. T

C. F

D. T

E. T Pallor of the dorsal motor nucleus is a feature

A3.46

A. T

B. T

C. F Hepatic cirrhosis stimulates growth hormone secretion

D. T

E. F Renal failure stimulates growth hormone secretion

A3.47
A. **F**   Cortical arousal is decreased
B. **T**
C. **F**   There is limited response to stress
D. **F**   Conditioning to fear-provoking stimuli occurs slowly
E. **T**

A3.48
A. **T**
B. **T**
C. **T**
D. **F**   Primary hypopituitarism stimulates prolactin secretion
E. **F**   There is no association

A3.49
A. **F**   The non-dominant parietal lobe is involved
B. **T**
C. **T**
D. **T**
E. **F**   Astereognosis in the left hand is caused by lesions in the corpus callosum

A3.50
A. **T**
B. **T**
C. **T**
D. **T**
E. **T**

A3.51
A. **T**
B. **F**   About 10% of total glucose turnover in the brain occurs via the GABA shunt
C. **T**
D. **F**   It requires glutamic acid decarboxylase
E. **T**

A3.52
A. **T**
B. **T**
C. **F**   DNA can be used as an indicator of protein/DNA ratio
D. **T**
E. **F**   Gangliosides can be used as indicators of brain specific protein S-100

A3.53

A. F SPECT (single photon emission computed tomography) employs radiochemicals that emit single photons
B. F The reverse is true
C. T
D. F There are small radiation risks
E. F This is not possible in SPECT; comparison is possible in PET

4. Psychopharmacology

Q4.1 The following are true about class I drug receptors:

A. They act via ion channels
B. They are slow acting
C. They act via the alpha unit of G protein
D. Muscarinic receptors are good examples of class I drug receptors
E. Nicotinic receptors are good examples of class I drug receptors

Q4.2 The following combinations of drugs are dangerous:

A. Fluoxetine and clomipramine
B. Buspirone and tranylcypromine
C. Lithium and fluvoxamine
D. Triazolam and trimipramine
E. Phenelzine and carbamazepine

Q4.3 Zopiclone:

A. Is a diazolobenzodiazepine
B. Binds to GABA receptors
C. Can be used in lactating mothers
D. Is effective in complex partial seizure
E. Does not suppress REM sleep

Q4.4 Lithium:

A. Inhibits extracellular phosphatase
B. Is indicated in refractory anxiety states
C. Increases adenylate cyclase activity
D. Levels are increased by carbonic anhydrase
E. Shows little variation in bioavailability with different preparations

Q4.5 For a drug whose elimination is first order:

A. The half-life increases as the dose administered increases
B. The elimination rate is constant
C. Linear kinetics is obeyed
D. The steady-state plasma level is proportional to the dose
E. The half-life is proportional to the plasma concentration

Q4.6   Following oral administration, drugs are absorbed:

A. Primarily by active transport
B. Mainly in the stomach
C. Better in the ionised form
D. Less readily in the presence of food
E. More slowly than when given by intramuscular injection

Q4.7   The following are true about opiates:

A. Opiate receptors are found in the thalamus
B. Opiates decrease respiratory depth
C. Opiates decrease the sensitivity of the respiratory centre to carbon dioxide
D. Opiates can be used as an antiemetic
E. Opiates decrease the respiratory rate

Q4.8   The following pairings of receptors and their antagonists are correct:

A. NMDA : dizocilpine
B. M1: pirenzepine
C. Alpha1: clonidine
D. Dopamine1: lisuride
E. $5HT_3$: raclopride

Q4.9   The following combinations of receptors and their agonists are correct:

A. $5HT_{1A}$: Ondansetrone
B. $5HT_{1B}$: RU-24969
C. $5HT_2$: ritanserin
D. Dopamine1 : SKF-38393
E. Dopamine2 : pergolide

Q4.10   The following combinations of drugs and mechanisms of action are correct:

A. Cocaine : inhibits noradrenaline (NA) reuptake
B. Cocaine : inhibits dopamine (DA) reuptake
C. Amphetamine : increases NA release
D. Amphetamine : increases DA release
E. Benztropine : inhibits DA reuptake

Q4.11 The following drugs are correctly paired with adverse effects on the foetus:

A. Diazepam : irritability
B. Diamorphine : floppy baby syndrome
C. Tricyclic antidepressants : tachycardia
D. Phenothiazine : congenital malformation
E. Lithium : Fallot's tetralogy

Q4.12 The following adverse effects result when the drugs are taken with disulphiram:

A. Antidepressants : decreased plasma concentration
B. Phenytoin : psychosis
C. Metronidazole : increased toxicity
D. Benzodiazepine : decreased sedation
E. Anticoagulants : increased bleeding tendencies

Q4.13 The following adverse effects result when the drugs are taken with monamine oxidase inhibitors (MAOIs):

A. Insulin : increased chances of hypoglycaemia
B. Oral antidiabetics : decreased effectivity
C. Anticonvulsants : decreased seizure threshold
D. Pethidine : hypertension
E. Tryptophan : confusion

Q4.14 The following adverse effects result when the drugs are taken with specific serotonin reuptake inhibitors (SSRIs):

A. Beta blockers : increased concentration
B. Clozapine : decreased concentration
C. Haloperidol : increased concentration
D. Tricyclic antidepressants : decreased concentration
E. Carbamazepine : increased concentration

Q4.15 The following are adverse effects of antiandrogenic drugs:

A. Fatigue
B. Breathlessness
C. Depression
D. Thrombosis
E. Hirsutism

4. Answers

A4.1
A. T
B. F   Class I receptors are fast acting
C. F   This is true for class II receptors
D. F
E. T

A4.2
A. T   This combination carries an increased risk of serotonin syndrome
B. T   Can cause hypertension
C. F
D. F
E. T   Can lower seizure threshold

A4.3
A. F   It is a cyclopyrrolone
B. T
C. F   It is excreted in breast milk
D. F
E. F   It does, but less than benzodiazepines

A4.4
A. F   Lithium inhibits intracellular phosphatase
B. F
C. F   It inhibits adenylate cyclase
D. F   Lithium levels are unaffected
E. F

A4.5
A. F   The half-life remains constant
B. F   The elimination rate is proportional to the plasma concentration
C. T
D. T
E. F   The half-life remains constant

A4.6
A. F Most drugs are passively absorbed
B. F They are absorbed mainly in the small intestine
C. F They are absorbed better in the non-ionised form
D. F The presence of food increases absorption of most psychotropic drugs
E. F Except for certain benzodiazepines, e.g. diazepam

A4.7
A. T
B. F Opiates decrease the rate, not the depth, of respiration
C. T But not to hypoxia
D. F They may cause vomiting
E. T

A4.8
A. T
B. T
C. F Clonidine is an agonist
D. T
E. T

A4.9
A. F It is an antagonist at $5HT_3$
B. T
C. F It is an antagonist at $5HT_2$
D. T
E. T

A4.10
A. T
B. T
C. T
D. T
E. T

A4.11
A. F It causes floppy baby syndrome
B. F It causes low birth weight
C. T
D. T
E. F It may cause Ebstein's anomaly

A4.12
A. F Increased plasma concentration results
B. F This combination causes increased toxicity
C. F This combination causes psychosis
D. F Increased sedation results
E. T

A4.13
A. T
B. T
C. T
D. F Hypotension results
E. T

A4.14
A. T Especially with fluvoxamine
B. F Increased concentration results
C. T
D. F Increased concentration results
E. T

A4.15
A. T
B. T
C. F Severe depression is a contraindication
D. F It is a contraindication
E. F It is an indication for use

5. Psychology

Q5.1   The following statements about the processes of learning of new behaviour are true:

A. Habituation is a complex form of learning
B. Conditioned stimulus is usually of biological significance
C. In classical conditioning subjects are passive
D. The main effect of operant conditioning is to increase the number of different stimuli to elicit a given response
E. In classical conditioning new behaviour can be learnt

Q5.2   The following are correctly paired:

A. Tolman : insight learning
B. Köhler : sign learning
C. Bandura : social learning
D. Thorndike : law of mass
E. Skinner : law of effect

Q5.3   Models from whom observational learning takes place have:

A. High status
B. High competence
C. High social power
D. High speed while speaking
E. Higher attractiveness

Q5.4   The following statements about different memory processes are correct:

A. Recency effect means that the first words learned are retained better
B. Episodic memory is subject to effort after meaning
C. Ribot's law states that retrograde amnesia affects more remote memories
D. The forgetting curve has a sharp initial gradient
E. The encoding specificity principle states that in certain cases recognition can be better than recall

Q5.5   Perceptual constancy has been demonstrated for:

A. Lightness
B. Location
C. Height
D. Colour
E. Depth

Q5.6   Perceptual set can be affected by:

A. Hunger
B. Thirst
C. Punishment
D. Personality
E. Experience

Q5.7   The following statements about perceptual threshold are correct:

A. Absolute threshold is tested by the method of descending limits
B. Large differences between two stimuli can be detected at low intensities, but only small differences can be detected at high intensities
C. The Weber–Fechner law measures the lowest intensity of a detectable stimulus
D. The Weber–Fechner law measures the smallest detectable change between two stimuli
E. The relationship between threshold of a stimulus and intensity is an inverted U shaped

Q5.8   The following statements about perceptual threshold are correct:

A. Absolute threshold is taken as the minimum amount of energy required to activate half the sensory organs
B. Difference threshold is the same as 'just noticeable difference'
C. Weber's law holds over a large range of stimulus intensities
D. Fechner's law states that sensory perception is a logarithmic function of stimulus intensities
E. Weber's constant for brightness of light is $\frac{1}{60}$

Q5.9   The following statements about learning theories are correct:

A. Theoretically, it is impossible to learn while under the influence of drugs
B. Cognitive learning is a type of associative learning
C. Delayed conditioning is optimal when the delay between two stimuli is about 0.5 second
D. Simultaneous conditioning is superior to delayed conditioning
E. In trace conditioning, the conditioned stimulus terminates before onset of the unconditioned stimulus

Q5.10   The following statements about learning theories are correct:

A. Thorndike is associated with operant behaviour
B. Skinner is associated with respondent behaviour
C. Operant behaviour is independent of stimulus
D. Respondent behaviour is independent of stimulus
E. Unconditioned responses in classical conditioning are a type of respondent behaviour

Q5.11   The following statements about operant conditioning are true:

A. Discrimination can occur in operant conditioning
B. The phenomenon of extinction is restricted to classical conditioning
C. A negative reinforcer reduces the probability of occurrence of an operant behaviour
D. Punishment is a type of negative reinforcement
E. In partial reinforcement, all the conditioned responses are partially reinforced

Q5.12   The following statements about the applications of associative learning are true:

A. Reciprocal inhibition is a psychological concept
B. Deep muscular relaxation is the most important element of systematic desensitisation
C. Exposure to the feared stimulus is the most important element of systematic desensitisation
D. A proper hierarchical presentation is the most important element of systematic desensitisation
E. Biofeedback is an application of classical conditioning

Q5.13 The following statements are true of non-associative learning:

A. Cognitive learning may occur suddenly
B. Insight learning is an active form of learning
C. In latent learning, learning is not manifested except in an emergency
D. Observational learning is a type of social learning
E. Social learning includes learning by classical conditioning

Q5.14 The following are true of human memory:

A. Olfactory information is stored as haptic memory
B. Short-term memory registers can be filled by parallel processing
C. 75% of information in short-term memory is forgotten by 9 seconds
D. The primacy effect does not hold true if more than seven items are presented for remembering
E. Verbal information in long-term memory is stored as words

Q5.15 The following are correctly paired:

A. Chaining : a method of successive approximation
B. Eysenck : preparedness
C. Mowrer : three-stage theory of phobia
D. Rachman : social skills training
E. Rutter : locus of control

Q5.16 The following statements about motivational theories are correct:

A. Multiple avoidance–avoidance conflicts are the most difficult to resolve
B. Love and belonging needs rank higher than self-esteem needs in Maslow's hierarchy
C. Intrinsic motivation theories suggest that needs arise to maintain biological homeostasis
D. Primary drives are always biological in nature
E. Achievement needs are concerned with the need to have impact, reputation and influence

Q5.17 The following statements regarding personality theories are correct:

A. In Kelly's theory, an individual is perceived as a philosopher
B. Eysenck's theory is a nomothetic type theory
C. Repertory grids can predict future behaviour
D. Roger's self theory suggests that there should be some incongruence between an individual's ideal and real self
E. Ideographic personality theories are based on a psychometric approach

Q5.18 The following statements are correct:

A. Variable interval reinforcement schedules take the longest to establish
B. The concept of incubation explains why some stimuli are more likely to condition than others
C. Perceptual constancy is fully developed by the age of 5 years
D. In competitive situations, an individual with an external locus of control does better than an individual with an internal locus of control
E. The concept of preparedness explains why phobic behaviour increases in severity

Q5.19 The following are correctly paired:

A. Perls : Gestalt therapy
B. Janov : primal therapy
C. Frankl : logotherapy
D. Binswanger : existential analysis
E. Sullivan : interpersonal psychotherapy

Q5.20 There is an inverse correlation between intelligence and:

A. Birth order
B. Family size
C. Age
D. Parental IQ
E. Marital status

Q5.21 The following associations are correct:

A. Spearman : S factor
B. Cattell : primary mental abilities
C. Hebb : type C intelligence
D. Thurstone : fluid ability
E. Vernon : type A intelligence

Q5.22 The following statements about memory types are correct:

A. Episodic and semantic are types of declarative memory
B. Semantic memory is essentially autobiographical in nature
C. In certain circumstances, it is easier to recall than to recognise information
D. Many amnesic patients have poor procedural but intact declarative memory
E. Traditional memory tests measure explicit memory

Q5.23 The following are correctly paired:

A. Adler : analytic psychology
B. Jung : individual psychology
C. Kohut : self psychology
D. Hartmann : ego psychology
E. Horney : basic need

Q5.24 The following statements about self-concept are true:

A. Physical attractiveness plays a prominent role in determination of self-esteem in children
B. Being popular is important in determination of self-esteem, especially in males
C. High self-esteem is associated with more risk-taking behaviour
D. Self-esteem is evaluative in nature
E. Self-image influences behaviour in a systematic and predictive manner

Q5.25 The following statements about stress responses are correct:

A. According to Seyle, most stress responses are non-specific
B. The level of body resistance is increased during the alarm reaction phase
C. Mineralocorticoids are released during the acute stress reaction phase
D. Platelet aggregation is reduced during the acute physiological stress reaction
E. Inflammatory responses are facilitated during acute physiological stress

5. Answers

A5.1
A. F It is the simplest form of learning
B. F Unconditioned stimulus is of biological significance
C. T
D. F This is true for classical conditioning
E. F This is true for operant conditioning

A5.2
A. F Tolman is associated with sign learning
B. F Köhler is associated with insight learning
C. T
D. F Thorndike is associated with law of effect
E. F New behaviours are learnt through operant conditioning

A5.3
A. T
B. T
C. T
D. F There is no association between speed of talking and observational learning
E. F There is no association between the attractiveness of the model and observational learning

A5.4
A. F This is primacy effect
B. T
C. F It affects more recent memories
D. T
E. F Normally recognition is better than recall but, in circumstances where recognition and retrieval contexts match, recall is better than recognition

A5.5
A. T
B. T
C. T
D. T
E. T

A5.6
A. T
B. T
C. T
D. T
E. T

A5.7
A. F Perceptual threshold is measured by the method of ascending limits
B. F It's the other way round
C. F It states that the relationship between perceptual threshold and intensity is a logarithmic one
D. F
E. F The relationship is logarithmic

A5.8
A. F Absolute threshold is the minimum energy required to activate a sensory organ in 50% of trials
B. T
C. F It fails to hold over a large range of stimulus intensity
D. T
E. T

A5.9
A. T
B. F Classical and operant conditioning are the two forms of associative learning; cognitive learning is a more complex process
C. T
D. F Delayed conditioning is superior
E. T

A5.10
A. T
B. T
C. T
D. F It is dependent on known stimuli
E. T

A5.11
A. T
B. F It can occur in respondent conditioning as well
C. F It is an aversive stimulus, the removal of which increases the prob-
ability of occurrence of an operant behaviour
D. F
E. F Only some of the responses are reinforced

A5.12
A. F It is a neurological concept
B. F
C. T
D. F It is not essential
E. F It is an application of operant conditioning

A5.13
A. T
B. F It occurs suddenly
C. F Learning occurs in response to the need to satisfy a basic drive
D. F It is a type of cognitive learning
E. T

A5.14
A. F Information from touch is stored as haptic memory
B. F Short-term memory registers are filled by the displacement principle
C. T
D. F It can hold true
E. F Verbal information is stored as meanings

A5.15
A. F It is a method of shaping
B. F Seligman is associated with preparedness
C. F Mowrer's is a two-stage theory
D. F Rachman is associated with the three-stage theory of phobia
E. T

A5.16
A. F Multiple approach–avoidance conflicts are the most difficult to resolve
B. F Self-esteem ranks higher than love and belonging needs
C. F This is true for extrinsic motivation theories
D. T
E. F Achievement needs are concerned with need to improve perform-
ance, to do better, etc.

A5.17
A. **F** In Kelly's theory, an individual is perceived as a scientist
B. **T**
C. **T**
D. **F** Roger's self theory suggests that there should be some congruence between real and ideal self
E. **F** Nomothetic theories are based on a psychometric approach

A5.18
A. **T**
B. **F** It explains why phobic behaviour increases in severity
C. **T** In some books 6–7 years is quoted as the limit
D. **F** Those with an internal locus of control do better
E. **F** It explains why some stimuli are more likely to condition than others

A5.19
A. **T**
B. **T**
C. **T**
D. **T**
E. **T**

A5.20
A. **T**
B. **T**
C. **F** Intelligence increases with age up to a certain limit
D. **F**
E. **F** There is no such relation

A5.21
A. **T**
B. **F** This is Thurstone's concept
C. **F** This is Vernon's concept
D. **F** This is Cattel's concept
E. **F** This is Hebb's concept

A5.22
A. **T**
B. **F** Episodic memory is autobiographical in nature
C. **T** According to Tulvig's encoding specificity principle
D. **F** Many amnesic patients have better procedural memory
E. **T**

A5.23
A. F Adler should be paired with individual psychology
B. F Jung should be paired with analytic psychology
C. T
D. T
E. T

A5.24
A. T
B. F It is more important in females
C. T
D. T
E. T

A5.25
A. T
B. F Decreased
C. T
D. F It is increased
E. F They are suppressed

6. Psychometry and Research Methodology

Q6.1 The following statements about IQ tests are correct:

A. The Stanford Binet and Wechsler scales correlate poorly
B. The upper IQ limits in Stanford Binet and Wechsler scales are the same
C. Mental age gives a 68% chance of correctly answering a question for the corresponding chronological age
D. Correlations between scores on Raven's Progressive Matrices and the Wechsler Adult intelligence scale (WAIS) are usually of the order of 0.75
E. The Raven's Progressive Matrices test does not involve recall of any learned information

Q6.2 Comparable IQ tests have shown that:

A. Boys are better at skills involving spatial relations
B. Girls are better at mathematics
C. Girls have higher IQ scores in childhood
D. Girls excel in vocabulary
E. Boys are more gifted

Q6.3 The following statements are correct:

A. Studies have shown that 50% of the variance in IQ of offspring is directly related to parental IQ
B. Modern IQ tests are remarkably exact at measuring intelligence
C. There is an inverse relationship between a child's IQ and the socio-economic status of the family
D. Childhood IQ is better at predicting adult educational attainment than adult IQ is
E. There is evidence that boys have a greater range of IQ than girls

Q6.4 The following are 'hold' tests:

A. Comprehension
B. Substitution
C. Vocabulary
D. Similarities
E. Digit span

Q6.5   The following are true about IQ tests:

A. IQ tests can specifically diagnose brain damage
B. In psychiatric illnesses in general, performance on verbal tests is less impaired than on non-verbal tests
C. Performance is better on verbal tests than on non-verbal tests in organic illness
D. Brain damage is more devastating in adults than in children
E. Old people are handicapped by reduction of 'fluid' intelligence

Q6.6   The following are true of personality tests:

A. The Q-sort technique compares the unique configuration of personality traits of different individuals
B. High scorers on the psychotism scale of the Eysenck Personality Questionnaire (EPQ) are claimed to resemble stereotyped psychopaths
C. The Minnesota Multiphasic Personality Inventory (MMPI) can measure defensiveness during answering the questionnaire
D. The EPQ incorporates a lie scale
E. The MMPI measures traits that are part of normal personality

Q6.7   The following statements about IQ tests are correct:

A. Mental ability in the Stanford Binet test is measured by mathematical and problem-solving abilities
B. In the Stanford Binet test, ten items are allocated to each year
C. The highest attainable score of chronological age in the Stanford Binet test is 15
D. Wechsler intelligence scales cannot measure the IQ of children below 5 years of age
E. DSM IIIR allows a measurement error of 10 points on IQ

Q6.8   The following are true statements about rating scales:

A. Forced choice techniques help to eliminate extreme forms of responding
B. Socially desirable responses occur only consciously
C. The phenomenon of social desirability occurs more commonly during interviews than in self-administered questionnaires
D. Lie scales detect deliberate liars
E. Response set occurs when subjects do not want to give away too much self-related information

Q6.9 The following statements about 'self' scales are correct:

A. Self-prediction is a valuable way of measuring behaviour
B. Self-prediction is most useful in alcoholics who are trying to reduce their drinking
C. Self-prediction is very useful to predict reoffending in prisoners
D. Self-recording is a better way of monitoring behaviour than self-prediction
E. Self-prediction is as useful as any other objective method of predicting relapse in smokers

Q6.10 The following statements about behavioural research are correct:

A. The halo effect is a common error made by subjects during interview
B. Video cameras should not be used during naturalistic observations
C. Naturalistic observations may be contaminated by Hawthorne effect
D. Time sampling is a type of naturalistic observation
E. Naturalistic observations can be used to carry out functional analysis of behaviour

Q6.11 The following statements about rating scales are correct:

A. The Likert scale is an equal interval scale
B. The reliability of an equal interval scale improves steadily up to a maximum of 5 points
C. When an investigator is measuring changes in the quality of an interpersonal relationship, it is important to use a generalisable scale
D. It is better to use an interval scale than a categorical scale when studying a psychological phenomenon in detail
E. A highly reliable scale may have poor validity, but a highly unreliable scale may be valid

Q6.12 The following statements about psychological research are true:

A. In any experiment, the more homogeneous a study sample is, the more generalisable the results are likely to be
B. None of the psychophysiological measures are unequivocally true
C. The eyeball approach is a useful tool to measure eye movement sensitivity in post-traumatic stress disorders
D. A rating scale records particular characteristics quantitatively
E. A rating scale can be called a questionnaire when statements are made in a question-and-answer form

Q6.13 The following statements about research methods are correct:

A. Data massaging in an experiment reduces the chances of type I errors
B. When many different raters are being assessed to measure reliability of an instrument, intra-rater reliability is more useful than inter-rater reliability
C. Product moment correlation coefficient is the most frequently used statistic to measure agreement between two raters
D. Product moment correlation coefficient is better than Kappa statistics in measuring the extent of agreement between two raters
E. Kappa statistics give a lower level of agreement than simple correlation measures

Q6.14 The following statements about hypothesis testing are correct:

A. Research in psychiatry is based primarily on proving rather than disproving a hypothesis
B. When a null hypothesis is disproved, it is automatically assumed that a directional hypothesis is proved
C. Recent evidence suggests that calculating an overall effect size for a hypothesis is probably better than doing costly time-consuming prospective studies
D. A pilot study is a small study to test the null hypothesis
E. A cohort study is especially useful in disorders for which no satisfactory treatment is available

Q6.15 In an epidemiological study:

A. A relative risk of 1 implies no causation
B. An attributable risk of 1 implies no causation
C. The relative risk can be calculated only in prospective studies
D. When a disease is relatively rare, an odds ratio can be calculated to check the relative risk
E. A positive attributable risk implies causation

Q6.16 The following statements are true of rating scales:

A. The General Health Questionnaire (GHQ) is particularly useful in identifying somatisation disorder
B. The Schedule for Affective Disorders and Schizophrenia (SADS) is a semi-structured interview schedule
C. The Present State Examination (PSE) can be used to determine caseness
D. SADS is more reliable than PSE in identifying schizophrenia
E. Case registers are useful for conducting cross-sectional studies

Q6.17 The following statements about caseness are correct:

A. Specificity of test is an index for caseness
B. Increasing the caseness threshold of a test increases its sensitivity
C. Increasing caseness threshold of a test decreases its specificity
D. The efficiency of a test is measured by the proportion of all the true results
E. To count as a case, standardised criteria require all the individual component thresholds to be passed

Q6.18 The following statements are true about distribution curves:

A. The arithmetic mean is a good measure of central tendency in a skewed distribution
B. The mode is greater than the mean in a positively skewed distribution
C. The mean is greater than the median in a positively skewed distribution
D. The mean is less than the mode in a negatively skewed distribution
E. The median is greater than the mode in a negatively skewed distribution

Q6.19 Standard deviation (SD):

A. SD is more difficult to calculate than quantile distribution
B. SD can have either a positive or a negative value
C. SD has the same units as the original observation
D. The variance is the square root of the SD
E. For a sample size of 15, a good estimate of population SD can be obtained by using 14 in the denominator of the equation

Q6.20 In statistical terms:

A. Events are said to be mutually exclusive when the occurrence of one, does not in any way influence the probability with which the other can occur
B. Events are said to be independent when the occurrence of one means that for all practical purposes, the other cannot occur
C. When events are independent, probabilities of one or the other occurring is the sum of the occurrence of their individual probabilities
D. When events are mutually exclusive, the probability that both will occur is equal to the product of their individual probabilities
E. The probability of an event occurring can have a maximum value of 1

Q6.21 The following statements about statistical distributions are true:

A. Poisson distribution is a type of binomial distribution
B. Binomial distribution is a type of continuous distribution
C. An F distribution is an asymmetrical distribution
D. A t distribution is a continuous distribution but with smaller tails than a normal distribution
E. A chi-squared distribution is an asymmetric distribution

Q6.22 The following statements about confidence intervals (CIs) are true:

A. CIs are a measure of hypothesis testing
B. A 95% CI in a study gives a probability of 5% of not including the estimated population parameters in that study
C. CIs help to decrease type I errors
D. A t distribution can be used to calculate CIs
E. CIs should be given for every comparison

Q6.23 The following statements about the null hypothesis are true:

A. It is a type of composite hypothesis
B. It is a type of simple hypothesis
C. It makes a statement diametrically opposite to the directional hypothesis
D. When rejected it gives the magnitude of difference between the study groups
E. It can be applied in chi-squared distributions

Q6.24 The following statements about non-parametric tests are true:

A. They are distribution free
B. They are easy to use as long as the sample size is more than 50
C. They lead to a higher probability of type I error
D. For a given sample size, they are more likely to detect a statistically significant result than a parametric test
E. A non-parametric equivalent of ANOVA is called the Kolmogorov–Smirnov test

Q6.25 The following statements about rating scales are true:

A. To imply that depression and anxiety are different points on a diagnostic scale, one would use an ordinal scale
B. Ordinal scales imply a hierarchy
C. On an interval scale, one can say that 20 feet is twice as long as 10 feet
D. No systematic relationship between different scores are implied on a categorical scale
E. A ratio scale is a type of interval scale

Q6.26 The following statements about PSE are correct:

A. It is a structured interview schedule
B. Each symptom is rated on a 5 point scale
C. It was originally designed for hospital in-patients
D. Items covering anxiety symptoms are more reliable than those covering depression
E. PSE can be used in the normal population

Q6.27 The following statements about SADS are correct:

A. It is a semi-structured interview schedule
B. There are three versions of it
C. It is mainly designed for use with hospital patients
D. It can diagnose patients with organic states
E. It cannot measure change

Q6.28 The following statements about Clinical Interview Schedules are correct:

A. It was devised by Spitzer in 1978
B. It is a partially structured interview schedule
C. It assesses symptoms in the last 1 month
D. It was designed for use with medical patients
E. It is recommend for use only by psychiatrists

Q6.29 The following statements about the GHQ are true:

A. It is a self-rated questionnaire of 60 items
B. Each question has five possible responses
C. It was designed for use in community settings
D. It can predict short-term response to various treatments
E. It has a scaled version

Q6.30 The following statements about the Brief Psychiatric Rating Scale (BPRS) are correct:

A. It contains 11 items
B. Each item is scored on a 5 point scale
C. It assesses symptoms in the last 4 weeks
D. It is unsuitable for use for patients with minor psychiatric illness
E. It is a structured interview schedule

Q6.31 The following statements about depression rating scales are true:

A. The Hamilton Rating Scale for Depression (HRSD) is a diagnostic instrument
B. The HRSD assesses symptoms in the last week
C. A score of 30 indicates severe depression on HRDS
D. The Beek Depression Inventory (BDI) is concerned exclusively with psychological symptoms of depression
E. The Montgomery and Asbeng Depression Rating Scale (MADRS) is particularly useful for assessing patients who are likely to experience marked side effects from medication

Q6.32 The following statements about the Diagnostic Interview Schedule are true:

A. It was devised by Robbins in 1981
B. It is a semi-structured interview schedule
C. It assesses symptoms that may have occurred at any time during the patient's life
D. It is possible to use it for organic disorders
E. Data collected from it are sufficient for making a diagnosis by Feighner's criteria

Q6.33 The following statements about anxiety rating scales are correct:

A. HAS was devised by Hamilton
B. HAD was devised by Snaith
C. The State–Trait and Anxiety Inventory (STAI) was devised by Spielberger
D. HAS is an adaptation of HAD
E. The Schedule for Affective Disorders and Schizophrenia–L (SADS-L) can assess anxiety

Q6.34 The following statements about mania rating scales are correct:

A. The Manic Rating Scale is rated on the basis of an 8-hour observation
B. It provides an objective measure of the severity of manic behaviour
C. It provides a measure of the frequency of manic behaviour
D. It is sensitive to change
E. It was designed for use by nurses

Q6.35 The following statements about obsessive–compulsive rating scales are correct:

A. The Leyton Inventory can produce scores for resistance
B. The Leyton Inventory can produce scores for interference
C. The Leyton Inventory can produce scores for traits
D. The Yale–Brown scale can measure only symptoms
E. The Maudsley Inventory is sensitive to change

Q6.36 The following are true of multivariate analysis:

A. It considers the relationship between a combination of two or more variables
B. Fundamentally, it involves manipulation of matrix data
C. In multiple regression, the predictor variable is being predicted from a linear combination of outcome variables
D. In linear discriminant analysis, the predictor variable has discrete levels
E. In MANOVA, the predictor variables are continuous and the outcome variables are discrete

Q6.37 The following are true of factor analysis:

A. It is used to study interrelationships among a set of variables without reference to a criterion
B. A matrix of correlations between every variable is created
C. The logarithmic combinations that best describe maximum correlation between the variables is called the principal component analysis
D. Each principal component is related to another according to a mathematical rule
E. Typically, only a few components are extracted in the initial study

Q6.38 The following are true of correlation coefficients:

A. The ϕ coefficient is used when both x and y variables are continuous
B. The ϕ coefficient is used when variable x is truly dichotomous and variable y is artificially dichotomous
C. The ϕ coefficient is used when variable x is artificially dichotomous and variable y is truly dichotomous
D. The biserial r coefficient is used when variables x and y are both continuous
E. The point biserial r coefficient is used when variables x and y are both artificially dichotomous

Q6.39 The following are true about ANOVA:

A. F distribution is the ratio of variances derived from two samples of the same population
B. It can be used to compare means of two or more samples
C. It is assumed that data from each sample should be normally distributed
D. In calculations of the F ratio, within sample variance occupies the numerator of the equation
E. The higher the value of the F ratio, the higher are the chances of rejecting the null hypothesis

Q6.40 The following methods can be used to correct confounding bias in an epidemiological study:

A. Restricting the study subjects
B. Increased matching of each case
C. Stratification of analysis
D. Multivariate techniques
E. Univariate analysis

6. Answers

A6.1
A. **F** There is a high degree of correlation between them
B. **F** The upper limits are different
C. **T**
D. **T** The range is 0.7–0.9
E. **T**

A6.2
A. **T**
B. **F**
C. **T**
D. **F** Girls have better linguistic ability, not necessarily vocabulary
E. **T**

A6.3
A. **F** About 25% of the total variance is explained by parental IQ
B. **F** Unfortunately they're not
C. **F** A direct relationship probably exists between IQ and socioeconomic status
D. **F** The reverse is true
E. **T**

A6.4
A. **T**
B. **F**
C. **F**
D. **T**
E. **F**

A6.5
A. **F** IQ tests can give a broad idea of brain damage
B. **T**
C. **T**
D. **F** The reverse is true
E. **T**

A6.6
A. F It captures unique configuration of traits within individuals
B. T
C. T It has a 30 item correction scale to measure how defensive a subject is in revealing his/her problems
D. T
E. F The California Psychological Inventory allows such measures

A6.7
A. F Mental ability is measured by level of problem solving and reasoning
B. F Six items are allocated to each year
C. T
D. F The Weschler Pre-school and Primary Scale of Intelligence can
E. F It allows a measurement of error of 5 points so that an IQ of 70 is considered to represent a band of 65–75

A6.8
A. F It helps in reducing socially desirable responses
B. F They also occur unconsciously
C. F The reverse is true
D. F Lie scales are included to reduce social desirability
E. F Response set measures defensiveness

A6.9
A. T
B. F It is valuable when former smokers are asked to predict the likelihood of resuming smoking
C. F It is of little value because they are strongly motivated not to be honest
D. F They are equally effective
E. T

A6.10
A. F It is an observer error
B. F They can be used
C. T
D. T
E. T

A6.11
A. T
B. F Reliability improves up to 7 points
C. F A simple analogous scale is enough
D. T
E. F Though the former is possible, the latter never is

A6.12

A. F Because the results will apply to a highly selected sample
B. F Most are, if carried out correctly
C. F It refers to assessment of any data subjectively, e.g. interpretation of EEG by simple visual examination of the traces
D. F It records qualitatively but measures quantitatively
E. F

A6.13

A. F It increases the chance of a type I error
B. T
C. T
D. F Kappa statistics are better
E. T

A6.14

A. F The reverse is true
B. F It merely becomes more acceptable than the null hypothesis until further information is available
C. F Effect size calculation only supports a hypothesis
D. F It tests the feasibility of a larger study
E. F It enables the study of the natural history of a disorder

A6.15

A. T
B. F A value of 0 implies no causation
C. F It may also be calculated in retrospective studies
D. T
E. F It implies association

A6.16

A. F
B. F It is structured
C. T By using the Index of Definition (ID)
D. F SADS can identify alcoholism and personality disorder, which PSE can't; there is no difference in reliability in identifying schizophrenia
E. F Case registers are useful for longitudinal studies

A6.17
A. F Sensitivity is an index of caseness
B. F Increasing the threshold increases specificity
C. F Increasing the threshold decreases sensitivity
D. T
E. F This may not be necessary, as a given condition may manifest itself in different ways

A6.18
A. F It is a good measure in a symmetrical distribution curve
B. F The mode is less than the median, which is less than the mean
C. T
D. T
E. F The mean is less than the median, which is less than the mode

A6.19
A. F
B. F SD is always positive
C. T
D. F Variance is SD squared
E. T

A6.20
A. F This is true for independent events
B. F This is true for mutually exclusive events
C. F This is true for mutually exclusive events
D. F This is true for independent events
E. T

A6.21
A. T
B. F Binomial distribution is a discrete distribution
C. T
D. F The tails are longer, but as the sample size increases it becomes similar to a normal distribution
E. T

A6.22
A. F It is a measure of estimation testing
B. T
C. F It helps to decrease type II errors
D. T
E. T

A6.23

A. F

B. T

C. T

D. F

E. T

A6.24

A. T

B. F The sample size must be less than 50

C. F They lead to a higher probability of type II error

D. T

E. F The Kruskal–Wallis test is a non-parametric equivalent

A6.25

A. F One would use a nominal test

B. T

C. F One can say this with a ratio scale

D. T

E. T

A6.26

A. F PSE is a semi-structured interview schedule

B. F A 3–4 point scale is used

C. T

D. F The reverse is true

E. T There is a 40 item version for use with a non-patient population

A6.27

A. F It is a structured schedule

B. T

C. T

D. F One cannot diagnose organic states using SADS

E. F It can measure change

A6.28

A. F It was devised by Goldberg in 1970

B. T

C. F CIS measures symptoms in the last week

D. F It is used in community survey

E. T

A6.29
A. T
B. F Each question has four possible responses: usual, no more than usual, rather more than usual, much more than usual
C. T
D. T
E. T

A6.30
A. F It contains 16 items: 11 by verbal report, five by observed behaviour
B. F Each item is scored on a 7 point scale
C. F There is no time limit
D. T
E. T

A6.31
A. F It is used on already diagnosed patients
B. F It assesses symptoms in the last few days
C. T
D. F MADRS is
E. T

A6.32
A. T
B. F It is highly structured
C. T
D. F
E. T

A6.33
A. F Snaith devised HAS
B. F Hamilton devised HAD
C. T
D. F HAS is an adaptation of CAS
E. T

A6.34
A. T
B. T
C. T
D. F No data are given
E. T

A6.35
A. T
B. T
C. T
D. T
E. T

A6.36
A. F Multivariate analysis considers the relationship between three or more variables
B. T
C. T
D. F Outcome variables have discrete levels; predictor variables are continuous
E. F In MANOVA, predictor variables are discrete and outcome variables are continuous

A6.37
A. T
B. T
C. F Principal component analysis uses linear combinations
D. F The principal components are independent of each other
E. T

A6.38
A. F The ϕ coefficient can be used for different types of variables, e.g. continuous, truly dichotomous and artificially dichotomous (when a continuous variable is converted into a dichotomous one) in various combinations, but not when both x and y are continuous
B. T
C. T
D. F Pearson r is used
E. F It is used when x is continuous and y is truly dichotomous

A6.39
A. T
B. F For comparing two samples, t tests are used; ANOVA is used for comparing three or more samples
C. T
D. F $F \text{ ratio} = \dfrac{\text{Between sample variance}}{\text{Within sample variance}}$
E. T

A6.40
A. T
B. T
C. T
D. T
E. F

7. Social Psychology

Q7.1 The following statements about personal relationships are correct:

A. Women with more attractive partners are less neurotic
B. Men with more attractive partners are less neurotic
C. Marital satisfaction is associated with sexual satisfaction
D. There is good evidence that complementarity (dominant with submissive partner) results in a more satisfactory marriage
E. Surveys of married adults have shown that overall happiness is most strongly related to satisfaction with work, income and leisure activities

Q7.2 The following statements about gender difference in behaviour of children are correct:

A. Boys are more aggressive (physically and verbally) than girls
B. Girls are more sociable than boys
C. Differences in level of aggression in boys and girls are greater in older children
D. Girls are more suggestible than boys
E. Boys have higher self-esteem and motivation to achieve than girls

Q7.3 The following statements about prosocial behaviour are correct:

A. Prosocial behaviour is defined as behaviour beneficial to the recipient but also of some benefit to the responder
B. Helping behaviour increases when rewarded and decreases when punished
C. Helping behaviour towards a stranger is reinforcing
D. You are more likely to help someone in an emergency when you have fellow adults around than if you are with small children
E. You are more likely to take positive action while witnessing a crime if you originated from a small town than from a big city

Q7.4   The following statements are true about altruistic behaviour:

A. Individuals high in need of approval are in general more altruistic than those low in this need, irrespective of whether their behaviour is being observed
B. Irrespective of your current need, you are always likely to help others if you have a strong belief in a just world
C. In various emergency situations, females receive help more often than males
D. In various emergency situations, females offer help more often than males
E. The 'just world' hypothesis was proposed by Lerner

Q7.5   The following statements about prejudice are true:

A. Prejudice is defined as a cluster of beliefs about minority group members
B. Discrimination is hostile feelings about minority group members
C. Bending over backwards to be friendly with a minority group member is a form of discrimination
D. Decreasing contact between different racial groups can reduce prejudice between them
E. Similarities in belief are more important than racial identity when two members of different races are involved in an intimate relationship

Q7.6   The following statements about persuasive communication are correct:

A. Rapid speakers are more persuasive than slow deliberate speakers
B. The rate of persuasion is directly proportional to the physical distance between the communicator and the recipient
C. The more distracted the recipients are, the smaller are their chances of being persuaded
D. Explicit messages are more persuasive for intelligent recipients
E. There is a U-shaped relationship between the anxiety level of recipients and the fear content of a message

Q7.7   In making social judgement:

A. People tend to over-use the available base rate information
B. People tend to pay most attention to the information that supports their preconceptions
C. People estimate fairly correctly how many others would give similar judgements
D. People are usually objective about their own expectations and beliefs
E. People overestimate the role of situational factors when judging others' behaviour

Q7.8   According to Hull:

A. The interpersonal distance appropriate between close friends is 1–3 feet
B. The interpersonal distance appropriate for impersonal contact is 4–10 feet
C. The interpersonal distance appropriate for physical sport is 0–1.5 feet
D. The interpersonal distance appropriate for business contact is 4–10 feet
E. The interpersonal distance appropriate for formal contact is 2–8 feet

Q7.9   The following statements about crowding are correct:

A. Social density is defined as the number of people in a given space
B. Spatial density is defined as the amount of space available for a given number of people
C. Social density decreases as the number of people increases in a given space
D. Spatial density increases as the number of people increases in a given space
E. Increasing social density adversely affects women more than men

Q7.10 The following statements about leadership styles are correct:

A. Autocratic style is more effective in Territorial Army (TA) training
B. Democratic leaders yield greater productivity from their group when a highly original product is required
C. Irrespective of intellectual ability, a directive leader improves the performance of a group
D. In the original experiment of Lewin, groups with democratic leaders performed better than those with autocratic leaders
E. In the absence of a leader, members in a *laissez-faire* group become aggressive to each other

Q7.11 The following statements about social facilitation of behaviour are correct:

A. The presence of others enhances performance on new and exciting tasks
B. Performance on tasks in front of others improves only if it is correct
C. Incorrect performances are corrected in the presence of others
D. Performance in front of others improve because we want to look good in others' eyes
E. Performance in front of others improves because of a conflict between paying attention to others and paying attention to the task in hand

Q7.12 The following statements about group decisions are correct:

A. Common sense suggests that decisions reached by groups are more conservative than those taken individually
B. Experimental findings suggests the opposite of the above
C. Group discussion causes members to shift towards a view which is opposite to the initial view
D. One of the reasons for group polarisation is the desire of each member to outdo each other
E. Being in a group makes people become more responsible for their actions

Q7.13 The following are true about the causes of aggression:

A. According to the drive theory, aggression stems from inner tendencies
B. Blocking of ongoing behaviour is thought to be the most powerful antecedent of aggression
C. Exposure to aggressive models has higher impact on aggression than physical attack
D. People who commit extreme acts of violence have weak inhibitions against aggression
E. The psychological character of empathy is known to reduce aggression

Q7.14 The following statements about social conformity are true:

A. Asch investigated the development of social norms using the autokinetic phenomenon
B. Conformity increases in a linear relationship with group size
C. Increasing social support from another group member increases conformity
D. Festinger proposed the social comparison theory to explain conformity
E. Most of us obey authority figures because they control powerful negative sanctions

Q7.15 The following statements about attitude and behaviour are correct:

A. Measured attitudes are poor predictors of behaviour
B. There is a high correlation between the individual measures of attitudes
C. Attitudes based on personal experience predict behaviour strongly
D. Global attitudes predict behaviour strongly
E. Attitudes are evaluative; beliefs are neutral

Q7.16 The following statements about scales for measuring attitude are correct:

A. A Thurstone scale is an equally appearing interval scale
B. A Likert scale is a ratio scale
C. Semantic Differential Scales control for positional responses
D. Bogardus formulated the Social Distance Scale to measure racial prejudice
E. Moreno coined the term *sociometry* to compare the attitudes of members working in a group

Q7.17 Cognitive dissonance can be reduced by:

A. Justification of effort
B. Changing attitudes
C. Taking stimulants
D. Rationalising the information creating the dissonance
E. Counter-attitudinal advocacy

Q7.18 High dissonance is likely to occur when there is:

A. High pressure to comply
B. Low perceived choice between two actions
C. Awareness of personal responsibility for the consequences of an action
D. An expected unpleasant consequence of the behaviour for others
E. Low arousal

Q7.19 Cognitive dissonance:

A. Is motivating
B. Leads to self-doubt
C. Leads to low self-esteem
D. Is ego syntonic
E. Explains behaviour markedly at variance with the initial attitude of an individual

Q7.20 The following statements about personal space are correct:

A. Young females interact at a closer distance with males than with other females
B. People with type A personality like close interaction
C. Females are concerned with invasion of personal space from the front
D. Males are concerned with invasion of personal space from the side
E. Violent people are concerned with invasion of personal space from the back

Q7.21 The following statements about individual behaviour are correct:

A. Heider proposed the balance theory
B. If you dislike someone very much, the fact that your beliefs do not match does not alter your cognitive balance
C. If you dislike someone, you would not pay any attention to their beliefs, except when they are of the same sex as you
D. If you disagree on a particular matter with someone you dislike, you won't question your own belief
E. To correct your own cognitive imbalance, you may alter your liking of a person, although your views may not change

Q7.22 The following statements about interpersonal attraction are correct:

A. Men prefer to pair with slightly less attractive women because there is less likelihood of rejection
B. Women seek men with a similar level of attractiveness
C. Men seek women who are more attractive than themselves
D. Women seek men who are less attractive than themselves
E. Physical attractiveness has no role to play in interpersonal attraction

Q7.23 The following coping mechanisms largely correspond with the respective defence mechanisms:

A. Objectivity : rationalisation
B. Concentration only on the task at hand : isolation
C. Suppression of inappropriate feelings : denial
D. Playfulness : regression
E. Empathy : reaction formation

Q7.24 The following statements about social facilitation of behaviour are correct:

A. Performance on new tasks is facilitated in the presence of members of the opposite sex
B. On additive tasks, individuals fare better than groups
C. On conjunctive tasks, groups fare better than individuals
D. Performance on disjunctive tasks is determined by the least competent worker
E. The strength of any social influence is diluted by the number of people receiving it

Q7.25 Vulnerability to group pressure is low in:

A. Single women
B. Protestants
C. White collar workers
D. Catholics
E. The unemployed

7. Answers

A7.1
A. **T**

B. **F** There is no such association as in women

C. **T**

D. **F** People who differ psychologically do not necessarily have more satisfactory marriages

E. **F** Overall happiness isn't; only marital happiness is related

A7.2
A. **T**

B. **F** There is no such evidence

C. **F** The differences are smaller in older children

D. **F** There is no such evidence

E. **F** There is no evidence

A7.3
A. **F** There is no obvious benefit for the responder

B. **T**

C. **T**

D. **F** The opposite is true: there is diffusion of responsibility when other adults are present

E. **T**

A7.4
A. **F** They are more altruistic only if their behaviour is observed

B. **F** You are likely to help others only if you are currently in need

C. **T**

D. **F** Males offer more help

E. **T**

A7.5
A. **F** Prejudice is defined as a cluster of negative beliefs

B. **F** Discrimination is negative action against a minority

C. **T** This is also known as reverse discrimination

D. **F** Increasing contact between different racial groups reduces prejudice

E. **F** Unfortunately racial identity is more important

A7.6
A. T
B. F An optimum distance is required
C. F More distraction leads to a higher chance of being persuaded
D. F They work better for less intelligent recipients
E. F The relationship is an inverted U shape

A7.7
A. F People under-use base rate information
B. T
C. F They overestimate this
D. F People are not objective; this phenomenon is called illusory correlation
E. F Dispositional factors are overestimated

A7.8
A. T
B. T
C. T
D. T
E. F It is greater than 12 feet

A7.9
A. T
B. T
C. F Social density increases
D. F Spatial density decreases
E. F It adversely affects men more than women

A7.10
A. T Not always though, especially when at war
B. F *Laissez-faire* groups yield greater productivity
C. F Intellectual ability does matter
D. F They were equal in performance
E. T

A7.11
A. F Performance is enhanced on well learned tasks only
B. T
C. F Incorrect responses are more impaired
D. T
E. T

A7.12
A. T
B. T
C. F The shift is in the same direction
D. T
E. F Deindividuation occurs in a group

A7.13
A. F This is the instinct theory of aggression
B. F It's only a weak determinant
C. F There is no such comparison; both are equally important
D. F Surprisingly, they have strong inhibitions against violence
E. T

A7.14
A. F Sherif used the autokinetic phenomenon
B. F Conformity does increase but only up to a certain limit
C. F It reduces conformity
D. T
E. F Not necessarily; we obey even when they don't control such sanctions

A7.15
A. T
B. F Individual measures do not correlate highly
C. T
D. F The more specific the attitude, the more predictable the behaviour
E. T

A7.16
A. T
B. F
C. F These scales do not control for positional response
D. T
E. T

A7.17
A. T
B. T
C. F
D. T
E. T Dissonance can be reduced by forced compliance

A7.18
A. F It occurs when there is low pressure to comply
B. F It occurs when there is high perceived choice
C. T
D. T
E. F It occurs when there is high arousal

A7.19
A. T
B. F
C. F There is no such relation
D. F It is ego dystonic
E. T

A7.20
A. T
B. F They need greater personal space
C. T
D. T
E. T

A7.21
A. T
B. T
C. F Sex has no role
D. T
E. T

A7.22
A. F
B. T Both men and women seek partners of a similar level of attractiveness
C. F
D. F
E. F

A7.23
A. F The corresponding defence mechanism is isolation
B. F The corresponding defence mechanism is isolation denial
C. F The corresponding defence mechanism is isolation repression
D. T
E. F The corresponding defence mechanism is isolation projection

A7.24

A. F Sex of observers has no role
B. F Groups fare better provided social loafing doesn't occur
C. F Individuals fare better; performance by groups is determined by the least competent worker
D. F It is determined by the most competent worker
E. T

A7.25

A. F
B. F
C. F
D. F
E. F

8. Social Sciences

Q8.1 The following statements are true of the sick role:

A. It was defined by Parson as a role taken by a sick individual depending on his or her own idea of what sickness is
B. It's abnormal
C. A person given a sick role has the right to reject appropriate help
D. The sick role is a way of legitimising illness behaviour
E. It's components become invalid in drug-induced psychosis

Q8.2 The following statements are true of illness behaviour:

A. It can be described in terms of stages
B. It's culturally determined
C. Contact with a doctor is necessary to legitimise illness
D. It can be abnormal
E. It can be equated with learned helplessness

Q8.3 The following statements are true about total institutions:

A. Goffman worked in St Elizabeth's Hospital in London
B. A large ship is a type of total institution
C. In a total institution, one needs to work for items to sustain life
D. The process by which an individual becomes institutionalised is called role stripping
E. Patients in an institution formed small colonies, a process Goffman termed colonisation
F. Wing used the term institutional neurosis to describe the withdrawn state of patients in a total institution
G. Wing coined the term secondary handicap
H. Secondary handicap is seen outside an institution
I. Secondary handicap results from the unfortunate way persons with primary handicaps react to themselves

Q8.4 The following statements are about social integration are true:

A. Communes are extreme forms of normal communities
B. Durkheim published his work on anomie in 1893
C. Hooliganism in a football match is an example of anomie
D. According to Durkheim, a healthy society has a wide ranging set of values
E. Disruption of collective conscience leads to anomie

Q8.5   In Creed's study of life events (LEs) and appendicitis:

A. Patients who experienced a severe LE in the preceding year had a higher incidence of inflamed appendix than those who did not
B. Depression rates were higher in patients with an inflamed appendix
C. Patients who experienced threatening LEs had a higher incidence of non-inflamed appendices than those who did not
D. Anxiety rates were higher in patients with a non-inflamed appendix
E. There was a causal relationship between LEs and appendicitis

Q8.6   The following are types of doctor–patient relationship as suggested by Szaz:

A. Activity–passivity
B. Passivity–activity
C. Cooperation–guidance
D. Guidance–cooperation
E. Mutual participation

Q8.7   In Rosenham's experiment:

A. The researcher and his co-workers admitted themselves by reporting paranoid delusions
B. It took them an average of 3 weeks to get discharged
C. After admission, they behaved as if hallucinated
D. Those who confessed to having been ill but said they were now feeling 'better' were kept in the longest
E. Those who professed their normality were discharged earlier

Q8.8   The following statements on Durkheim's work on suicide are correct:

A. Modern studies of suicide have proved Durkheim's earlier work to be incorrect
B. Insights from Durkheim's work has been extended to other forms of illness
C. In Durkheim's work, Protestants showed a higher suicide rate than Jews
D. Durkheim argued that pre-industrial societies were characterised by high social integration
E. Durkheim argued that fatalistic suicide was produced by loose social regulation

Q8.9   Studies of sickness-absence rates have shown that:

A. They are valid measures of health and illness
B. They are affected by sick role
C. They include the unemployed
D. They measure the amount of sickness in the healthiest group of the population
E. They are limited measures of health status

Q8.10   The following are true about activities of daily living (ADL):

A. Katz formulated the ADL measures
B. ADL measures the major everyday functions only
C. Having a bath is measured as a major function
D. The responses are added to give an overall score of abilities
E. Doing up zips and buttons is considered to be a major activity

Q8.11   Studies of illness behaviour have shown that:

A. Rates of medical consultation accurately reflect the frequency of symptoms experienced
B. Older people with aches and pain consult their doctors more frequently than younger ones
C. Normalisation of some symptoms blocks the emergence of important diagnostic information
D. Most patients go to doctors immediately after they perceive a symptom
E. Very few of the symptoms perceived are self-limiting
F. Seeking medical help is almost always related to severity of illness
G. Patients with known heart disease go to doctors quicker than those without when they experience chest pain

Q8.12   In traditional culture, sick roles are:

A. Universal
B. Diffuse
C. Ascriptive
D. Particular
E. Achieved

Q8.13   The following help to break down the sick role:

A. An increasing incidence of chronic conditions
B. Medicalisation of social problems
C. Ageing of the population
D. Self-help movements
E. Renewed interest in preventive medicine

Q8.14 The following statements are true about the medical profession:

A. It enjoys greater autonomy than other professions
B. It is more likely to be a terminal occupation
C. It is relatively free of lay evaluation
D. Student doctors go through more liberal socialisation than other students
E. Most legislation concerned with the profession is shaped by legal authorities

Q8.15 Morbidity studies have shown that:

A. Married men have higher rates of strokes than single men
B. Single men have lower rates of heart disease than married men
C. Married women are more at risk of developing mental illness than single women
D. Despite changes in the social role of women, the overall morbidity figures remain the same
E. Married men have higher rates of lung cancer than single men

Q8.16 According to the OPCS data (1994):

A. The average household size in the UK is 2.45
B. Lone-parent families make up 10% of all families
C. One-person families make up 10% of all families
D. Typical families make up 40% of all families
E. Twenty-three per cent of families are couples with no children

Q8.17 The following are risk factors for marital breakdown:

A. Childlessness
B. Marriage for ten or more years
C. Both partners being younger than 20 years at the time of marriage
D. The bride being pregnant at the time of marriage
E. Having five or more children under the age of 11

Q8.18 The following statements about social class are true:

A. Social class is the same as social status
B. Higher class people are superior to lower class people
C. Lower class people believe in a social hierarchy
D. Middle class people believe in social mobility
E. In a crisis, middle class people stick together more readily than lower class people
F. Middle class people sympathise with those needing social assistance more readily than lower class people

Q8.19 The following statements are correct:

A. Working class people consult health services more frequently than middle class people
B. Working class people have higher availability of health resources than middle class people
C. Working class people have greater health needs than middle class people
D. Working class people consult their doctors more often for sickness certificates than middle class people
E. Availability of good medical care varies directly with the health needs of the population

Q8.20 Surveys have shown that:

A. Middle class patients have shorter consultations with their doctors than working class patients
B. Families of unskilled workers are more likely to have strong role segregation between husband and wife
C. Families of unskilled workers are more likely to stereotype children in terms of gender
D. The risk of death after retirement is twice as high in social class 5 as in social class1
E. Rates of longstanding illness are twice as high in unskilled workers as in professionals

Q8.21 According to Rack, *Gastarbeiter*s are:

A. People from developing countries who migrate to developed counties
B. Middle-aged men
C. Migrating men with no intentions of returning to their countries of origin
D. Migrating families seeking better opportunities
E. Rural people leaving the countryside for economic reasons

Q8.22 Jarmen indices:

A. Are measures of social deprivation
B. Were devised by the sociologist Norman Jarmen
C. Are devised from data derived from epidemiological surveys
D. Include the elderly living alone
E. Include mobility

Q8.23 The following are true about illness and social causation:

A. Japanese men who emigrate to the USA are more likely to die of stroke than those who stay in Japan, where ischaemic heart disease (IHD) causes more deaths
B. American men are more likely to die from stroke than from IHD
C. People in social classes 1 and 2 have a higher mortality rate from IHD than those in classes 4 and 5
D. Compared to others in the UK, there is a higher rate of IHD among Asians in the UK
E. Compared to others in the UK, there is a higher rate of hypertension and stroke among Afro-Caribbeans in the UK

Q8.24 The following are true about life events and mental illness:

A. Studies involving the Schedule of Recent Events (SRE) have shown that as many as one-third of events are missed
B. The fall off rate of the Life-Events and Difficulties Schedule (LEDS) is around 5%
C. The SRE uses the concept of life change units
D. The LEDS uses a 4 point scale of threat
E. The SRE scoring system assumes that the effects of life events are additive

Q8.25 The following are correct:

A. The relationship between increase in the number of life events and relapse of schizophrenia holds true only for severe events
B. The relationship between increase in the number of life events and relapse of depression holds true for all types of event
C. The brought-forward time for depression is about 9 months
D. The brought-forward time for schizophrenia is about 3 weeks
E. There is a consensus that severely threatening life events are associated with the onset of organic disorders only when there is concurrent psychiatric disorder

8. Answers

A8.1
A. **F** The sick role is given by the society
B. **F**
C. **F**
D. **F** Doctors legitimise illness
E. **T**

A8.2
A. **T**
B. **T**
C. **F**
D. **T**
E. **T**

A8.3
A. **F** St Elizabeth's is in Washington
B. **T**
C. **F** Items are automatically provided
D. **F** It is called the mortification process
E. **F** Colonisation was a process whereby the patients pretended to show acceptance of institutionalisation
F. **F** Barton coined the term institutional neurosis
G. **T**
H. **T**
I. **F** Secondary handicap results from others' reactions

A8.4
A. **F**
B. **T**
C. **F**
D. **F** Durkheim felt that a healthy society has only a few common sets of values
E. **T**

A8.5
A. **F** These patients were more likely to have non-inflamed appendices
B. **F** Patients with a non-inflamed appendix had higher depression rates
C. **F** They had a higher incidence of pathologically inflamed appendix
D. **F** This was not studied
E. **F** Only an association was found

A8.6
A. T
B. F This was suggested by Friedson
C. F
D. T
E. T

A8.7
A. F They reported hallucination the night before
B. F They were discharged, on average, after 30 days
C. F They behaved completely normally
D. F They were discharged earlier
E. F They were kept longer

A8.8
A. F
B. T For example, the influence of social integration on health
C. T Because of lower social integration
D. T
E. F Fatalistic suicide was caused by over-regulation

A8.9
A. F
B. F They are affected by illness behaviour
C. F
D. T
E. T

A8.10
A. T
B. F it also measures minor functions
C. F This is classed as a minor function
D. T
E. T

A8.11
A. F Several other factors are considered before going to doctors
B. F Because aches and pains are considered normal
C. T
D. F
E. F Most of them are
F. F
G. F

A8.12
A. F They are particular
B. T
C. T
D. T
E. F They are ascribed

A8.13
A. T
B. T
C. T
D. T
E. T

A8.14
A. F
B. T Once a doctor, a person is likely to continue to be one
C. T
D. F Their socialisation is more stringent
E. F It is shaped by the profession

A8.15
A. F Married men have lower rates of stroke
B. F Single men have higher rates
C. T
D. T
E. F Married men have lower rates

A8.16
A. T
B. T
C. T
D. T
E. T

A8.17
A. T
B. T
C. T
D. T
E. F

A8.18
A. F Class is determined more by life chances, attitudes and values, whereas status is determined more by occupation, skills, other behaviour, etc.
B. F Only patterns of behaviour differ
C. F They believe that the society is divided into 'us' and 'them'
D. T They believe that mobility depends on utilisation of individual abilities
E. F That's a lower class perspective
F. F They do not; lower class people do

A8.19
A. F Working people consult less frequently
B. F
C. T
D. T
E. F It varies inversely

A8.20
A. F Their consultations are longer
B. T
C. T
D. T
E. T

A8.21
A. F
B. F They are young men
C. F These are called settlers
D. F
E. T

A8.22
A. T
B. F Brian German, a GP
C. F They are devised from census data
D. T
E. T

A8.23
A. F

B. F More Japanese men in Japan die from stroke and more American men in America die from IHD, but more Japanese men in America die from IHD

C. F People from lower classes have lower mortality rates from IHD

D. T

E. T

A8.24
A. T

B. F It is 1%; it is 5% for SRE

C. T

D. T

E. T

A8.25
A. F The relationship holds for any type of life event

B. F The relationship holds only for severe events

C. F It's about 1–2 years

D. F It's about 10 weeks

E. T

Bibliography and References

1. Caplan, H. I. and Saddock, B. J. (Eds) (1995). *Comprehensive Textbook of Psychiatry*, Vol. I, 6th edn, Williams & Wilkins, Baltimore.
2. Snell, R. S. (1987). *Clinical Neuroanatomy for Medical Students*, 2nd edn, Little Brown, Boston.
3. Morgan, G. and Butler, S. (Eds) (1993). *Seminars in Basic Neurosciences*, Gaskell, London.
4. Trimble, M. E. (1996). *Biological Psychiatry*, 2nd edn, Wiley, Chichester.
5. Lader, M. and Herrinton, R. (1990). *Biological Treatments in Psychiatry*, OUP, Oxford.
6. Puri, B. K. and Tyrer, P. J. (1992). *Sciences Basic to Psychiatry*, Churchill Livingstone, Edinburgh.
7. Weller, M. and Eyesenck, M. (Eds) (1992). *The Scientific Basis of Psychiatry*, W. B. Saunders, London.
8. Freeman, C. and Tyrer, P. (Eds) (1992). *Research Methods in Psychiatry: A Beginner's Guide*, 2nd edn, Gaskell, London.
9. Henderson, A. S. (1990). *An Introduction to Social Psychiatry*, OUP, Oxford.
10. Armstrong, D. (1994). *Outlines of Sociology as Applied to Medicine*, 4th edn, Butterworth-Heinemann, Oxford.
11. Tantum, D. and Birchwood, M. (Eds) (1994). *Seminars in Psychology and Social Sciences*, Gaskell, London.
12. Byron, R. A. and Byrne, D. (1977). *Social Psychology: An Understanding of Human Interactions*, 2nd edn, Allyn and Bacon, Boston, Massachusetts.
13. Atkinson, R. I., Atkinson, R. C., *et al.* (1993). *Introduction to Psychology*, 11th edn, Harcourt Brace, London.
14. Kendell, R. E. and Zealley, A. K. (1994). *Companion to Psychiatric Studies*, 5th edn, Churchill Livingstone, Edinburgh.
15. *British National Formulary*, No. 33, March 1997.

Printed and bound by CPI Group (UK) Ltd, Croydon, CR0 4YY

23/10/2024

01777665-0008